NORTHERN ILLINOIS FIRE GROUND PHOTOGRAPHY

By Fire Department Photo Unit, Tim Olk

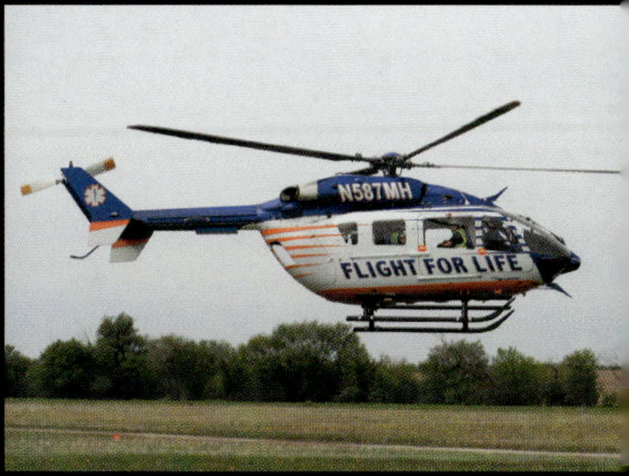

M.T. Publishing Company, Inc.™

P.O. Box 6802
Evansville, Indiana 47719-6802
www.mtpublishing.com

Library of Congress Control Number: 2013936915

ISBN: 978-1-938730-03-0

Design By:
Thalita A. Floyd-Wingerter
M.T. Publishing Company, Inc.™

First Printing July 2013
Second Printing December 2013

Printed in the
U.S.A.

This Book Is Dedicated To All My Friends Who Have Passed On Who Helped Me Out Sometime In My Life

NorthbrookAssistant Chief Wayne Luecht

Chicago Fire Department Lieutenant Pat Hannon

Northbrook Fire Department Firefighter Dave Rivest

Chicago Fire Department Captain Herbie Johnson

Chicago Fire Department Firefighter Chris Wheatley

Chicago Fire Department Firefighter Billy Grant

Chicago Fire Department Retired Chief Ray Hoff

All the 343 Members of the FDNY who were killed on 9-11

Gurnee Fire Department Firefighter Russ Constantino

Lake Forest Fire Department Firefighter Pete McWilliams

My dad, Jerry Olk

Background image: Lake County Fire Chiefs Fire Expo

Fire instructors from the Illinois Fire Service Institute (IFSI) teach propane fires to the Firefighter 2 Academy participants at the Northeastern Illinois Public SafetyTraining Academy (NIPSTA).
(March 2009)

The Author & Photographer

Tim Olk took his first fire scene photos when he was 13 years old. He brought the pictures into the station and met Assistant Chief/Fire Investigator Wayne Luecht. Tim started exploring departments in other nearby cities and his interest grew. He took photography classes and learned how to shoot fire scenes from Luecht.

Eventually, Tim became a Paid on Call Firefighter as well as certified in Advanced Fire Scene Photography by the Fire Investigators Strike Force. Tim is now the Midwest U.S. Region Director for the International Fire Photographers Association. He has had many photos featured in *Fire Engineering*, *Fire Chief* and *Fire Rescue* magazines. In 1999, Tim lost his mentor and friend when Wayne Luecht died of injuries sustained on duty.

Tim is still traveling all over the greater Chicago area capturing scenes like the many shared in this book.

Acknowledgments

Large Thanks Go Out To ~

My wife Jill.

My parents, Jerry and Margaret Olk; my brother, Northfield Fire Captain Tom Olk; my sister, Theresa Peiffer; and my mother-in-law and father-in-law John & Jean Dreckmeier.

Chicago Fire Department Lieutenant Pat Hannon, his wife Chicago Firefighter Michelle Hannon, and their daughters Katie and Bridget.

Lyons Fire Department Fire Chief Gordon Nord; Chicago Fire Department Fire Commissioner Jose Santiago; Former Fire Commissioners Bob Hoff and Ray Orozco; Carol Stream Fire Chief Rick Kolomay; Milwaukee Fire Department Battalion Chief Erich Roden; *Fire Engineering* magazine's Bobby Halton; Newport Fire Department Chief Mark Kirschhoffer; Round Lake Fire Department Fire Chief Mapplethorpe; Gurnee Firefighter Paramedic John Rikje; and Illinois Mutual Aid Box Alarm System (MABAS) CEO Jay Reardon; Chicago Squad 1, First Shift Lieutenant William "Mother Duck" Oswald, Firefighter Robert "Hoss" McCrea, Firefighter/Paramedic Patrick "Tango" Brown, Firefighter/Paramedic Henry "Just Henry" Hugel, Firefighter/EMT Michael "Angry Bird" Cronin, Firefighter/EMT Eric "Angry Inch" Sobolewski, Firefighter/EMT Michael "Eagle Eye" Chesack, Firefighter/EMT Lydara "Sweet Tea" Turner, Firefighter/EMT Craig "Otis" Dziedzic, Firefighter/EMT Clarence "Hollywood" Norwood, and Firefighter/Paramedic Joe "Goose Egg" Pullano.

Chicago Fire Department Chief John McNicholas, Chief Fox, Chief Edgeworth, Chief Bobby McKee, Chief Callahan, Chief Sampey, Chief Chikerotis, Chief Witt, Chief Niego, Chief Nielson, Chief Springer, Chief Hroma, and Chief Lee Basile; The Bull; Chicago Fire Department Squad 1, CFD Squad 2, CFD Squad 5 and CFD Squad 7, and all the members who have helped me out through the years at fire scenes.

Chicago Fire Department Paramedic Brandy Dooley; Chicago Fire Department Paramedic Kelly Begley; Schaumburg Firefighter Jenna Werdell; Rich Pinsky from Ignite the Spirit; Tim O'Brian; Father Tom Mulcrone; Southside Fraternal Order of Leatherheads Society (FOOLS); Tom Ryan, CFD Local 2; Chief Pete Vandorpe; Chief Gordon Nord; Chief Bill Fredrick; Lake Forest Fire Chief Howell; Countryside Chief Jeff Steingart; Long Grove Fire Department Chief Turpel; Lake Forest Deputy Chief Chris Garrison; Northbrook Chief Mark Nolan; Stillman Valley Deputy Chief Greg Kunce; Lake Forest Deputy Chief Kevin Issel; Retired Knollwood Chief Chuck Knesley; Retired Northbrook Deputy Chief Bob Brogan; Retired Battalion Chief Pat Crown; Retired Northbrook Battalion Chief Jim Hoover; Retired Northbrook Captain Bernie Schmidt; Glenview Fire Chief Wayne Globerger; North Maine Battalion Chief Mike Fox; Retired Franklin Park Fire Chief Dave Traiforos; Terry Cox, Mess Canteen; 5-11 Club; Mark Typlin, Fire Paging; Red Center MABAS Division 3; Larry Shapiro; Steve Redick; Chicago Fire Department Firefighter Kelly Burns; Chicago Fire Department Lieutenant Corey Hojek; Chicago Fire Department Chief Mitch Crooker; Chicago Fire Department Special Operations Chiefs Jim Purl, John Collins and Pat Maloney; NIPSTA Fire Acdemy Instructors; Chicago Fire Department Retired Squad 2 Lieutenant Pat Lynch; all MABAS Division 4 Departments and MABAS Division 3 Departments; Round Lake Firefighter Mike Bloom; Chicago Fire Department Office of Fire Investigations; State Fire Marshall K-9 Mitch and K-9 Radar; Tammy Chatman, Flight for Life; Frankfort Firefighter Cathy Crook; Fire Department Instructors Conference (FDIC) Instructors,

Northbrook Fire Department Memorial outside Station 11 for Assistant Chief Wayne Luecht, who was critically burned in a fire at Northbrook Court Mall in June 1999.

2004

Chicago Fire Department Retired Paramedic Field Chief 453 Mike Flaherty on his last day

December 2012

Chicago Fire Department Squad 1 members after a still alarm fire with Battalion 3 Chief Lee Basile—Lieutenant Maibuecher, Firefighter Kelly Burns, Firefighter Jeff Weber, Firefighter Frank Kaufman, Firefighter Jim Sewnig and Firefighter Eric Troken.

Highland Park firefighters at an extra alarm house fire in Lake Bluff on New Year's Day—Firefighter Stanley, Lieutenant Kwasny, Firefighter Salazar and Firefighter Maslon.

Glenview Fire Department Chief Wayne Globerger and son Kurt

Chicago Fire Department
Squad 2 Firefighter
Art Aranda
(2012)

FDNY Lieutenant and
Instructor Mike Ciampo
(April 2011)

Franklin Park
Retired Chief
Dave Traiforos
(2011)

Chicago Fire Department District Chief
Ken Wojtecki, Chief of Fire Safety
(2009)

Chicago Fire Department
Chief Kevin Krasneck
(2011)

July 2009

July 2009

Buffalo Grove firefighters put out a plane engine fire after the plane crashed into a field. Two people on board suffered minor injuries.

PAGES 10 &11, PHOTOS AT TOP: Somers, Wisconsin Fire Department performs a 15-hour trench rescue recovery of a construction worker who was buried 25 to 30 feet below ground.

Highland Park Firefighter Ryan Hughes
(July 2007)

Leyden firefighters search several houses after heavy rains flooded the town.
(September 2008)

Glenview Fire Department Lieutenant Tony Angus (June 2004)

Detroit Firefighter Steve Florian and FDNY Lieutenant Ray McCormack teach at the Fire Department Instructors Conference (FDIC) in Indianapolis. (April 2011)

Chicago Fire Department live fire training at the old Robert Taylor Homes at 51st and Federal; from left, Firefighter Infante, Chief Witt, Firefighter Sewnig, Firefighter Lynch and Firefighter Dave Owsianiak

November 2006

NIPSTA Fire Academy live car fires

Antioch Fire Department divers search for a person who fell into the water at Grass Lake and never resurfaced. (2013)

Woodstock Fire Department conducts live fire training on a house that was donated to them. (2013)

Barrington firefighters pose in front of their ladder at the scene of a fully involved vacant barn fire just off Penny Road.
(2013)

Wilmette firefighters cut free six people from a car that a tree fell on while they were driving at 16th & Willmette Avenue.
(2013)

Wilmette firefighters get ready to attack a second floor bedroom fire in a house off Old Glenview Road.
(2009)

Broadview Deputy Chief Jack Kowalski and Park Ridge Dive Master Russ Lococo

September 2008

Chicago Fire Department Squad 1 at the scene of an L-train fire—Firefighter Dave Leon, Captain John Collins, Firefighter Jim Stepien, Firefighter Kelly Burns, Firefighter Corey Hojek and Firefighter Brandon Dyer

BELOW: Forest Park Fire Department box alarm fire at 7230 W. Madison Avenue on September 2, 2011.

Lake Forest Fire Department at a 2nd alarm house fire at 1590 Waukegan Road. First-in companies had very long lead-outs due to the house being about a mile off the road.

LEFT: Elk Grove Township firefighters work a 2nd alarm trailer park fire off Touhy Avenue; (ABOVE) Engine 11 pumps at the Touhy Avenue fire.

McHenry County College Firefighter 2 live fires in Wauconda.

2010

Waukegan Lieutenant Henry Gruba watches the McHenry County College Firefighter 2 live fires in Wauconda.

Ron Moritz and his son help out with the McHenry County College Fire Academy.

Chicago Fire Department firefighters from Squad 5 and Engine 116 at a 2-11 alarm fire at 48th & Halstead (2013)

Chicago firefighters and Glenview firefighters work to put out a basement fire in Lincolnwood. (2011)

Chicago Fire Department Battalion Chief climbs a fire escape to check for an access point to an extra alarm fire at the Chinese museum in Chinatown. (2008)

TOP RIGHT: Cary Fire Department Firefighter Lance Neuse pulls an exposure line at live fire training in Cary, Illinois.

CENTER RIGHT: Des Plaines Fire Department Tower Ladder 81 works at a 2nd alarm commercial structure fire at 2501 E. Oakton in Elk Grove Township.

BOTTOM RIGHT: Elk Grove Township firefighters do salvage and overhaul at a 2nd alarm structure fire at 2501 E. Oakton.

Bedford Park Fire Department tower ladder knocks down a fire that was in two freight cars filled with cardboard.

2012

2007

Newport Chief Mark Kirschhoffer teaches roof operations to a couple of Beach Park firefighters on one of several training burns in a house set to be torn down for the Waukegan airport. (2007)

CENTER LEFT: Newport firefighters train on a house in Beach Park; this is one of several houses that were being torn down for the Waukegan airport.

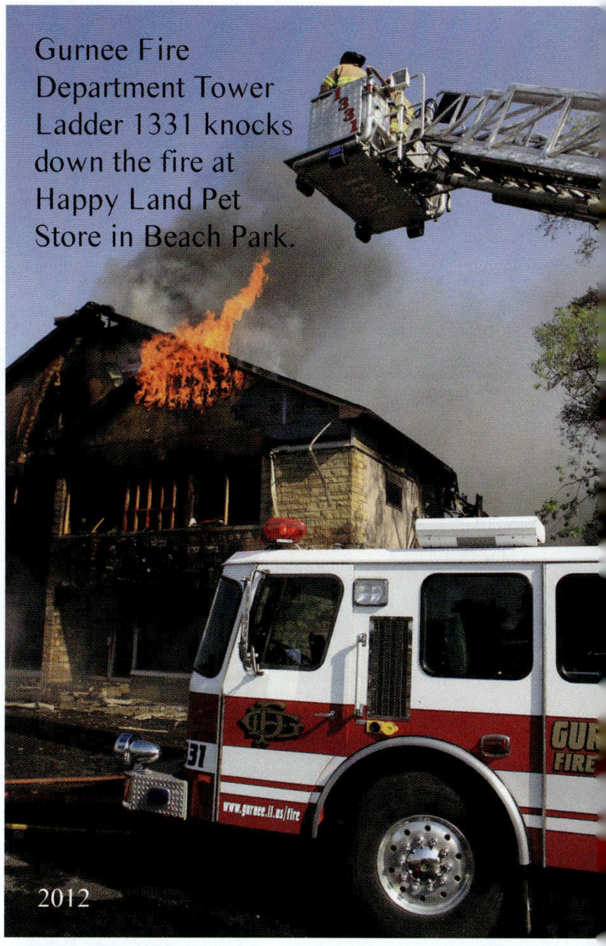

Gurnee Fire Department Tower Ladder 1331 knocks down the fire at Happy Land Pet Store in Beach Park.

2012

Two Countryside firefighters open a door at the Happy Land Pet Store fire in Beach Park.

2012

2012

2012

PHOTOS ON PAGE 20 & ON PAGE 21, ABOVE: Chicago Fire Department Tower Ladder 39 at the funeral of Captain Herbie Johnson (of Engine 123 and Tower Ladder 39), who was killed in the line of duty while fighting a fire on the southwest side.

CFD retired member Paul Enhelder

On Page 22—

Top right: Chicago Fire Department Squad 2 firefighters at the Marine Reserve Family Day

Center: Lake Zurich firefighters open up the garage doors at the Barrington Countryside 4-11 alarm fire at 21089 Ridge Road.

Bottom: Chicago Heights Fire Department box alarm fire in a vacant two-story building in the 200 block of U.S. 30 on December 24, 2010.

On Page 23—

Top left: Northfield Fire-Rescue Captain Tom Olk instructs his guys where to cut the vent hole at a Code 4 house fire at 4th & Maple in Wilmette.

Center: Chicago firefighters go defensive on a 1½-story house at 1416 W. 112th Place; three firefighters were hurt in a flashover.

Bottom: Antioch firefighters take time to pose for a picture after they were promoted.

Chicago FD Squad 2 Firefighter Bryan Velez

Top: Chicago FD Squad 2 Firefighter Ken Jurasz; (bottom) CFD Squad 2 Firefighter Clay Whitley

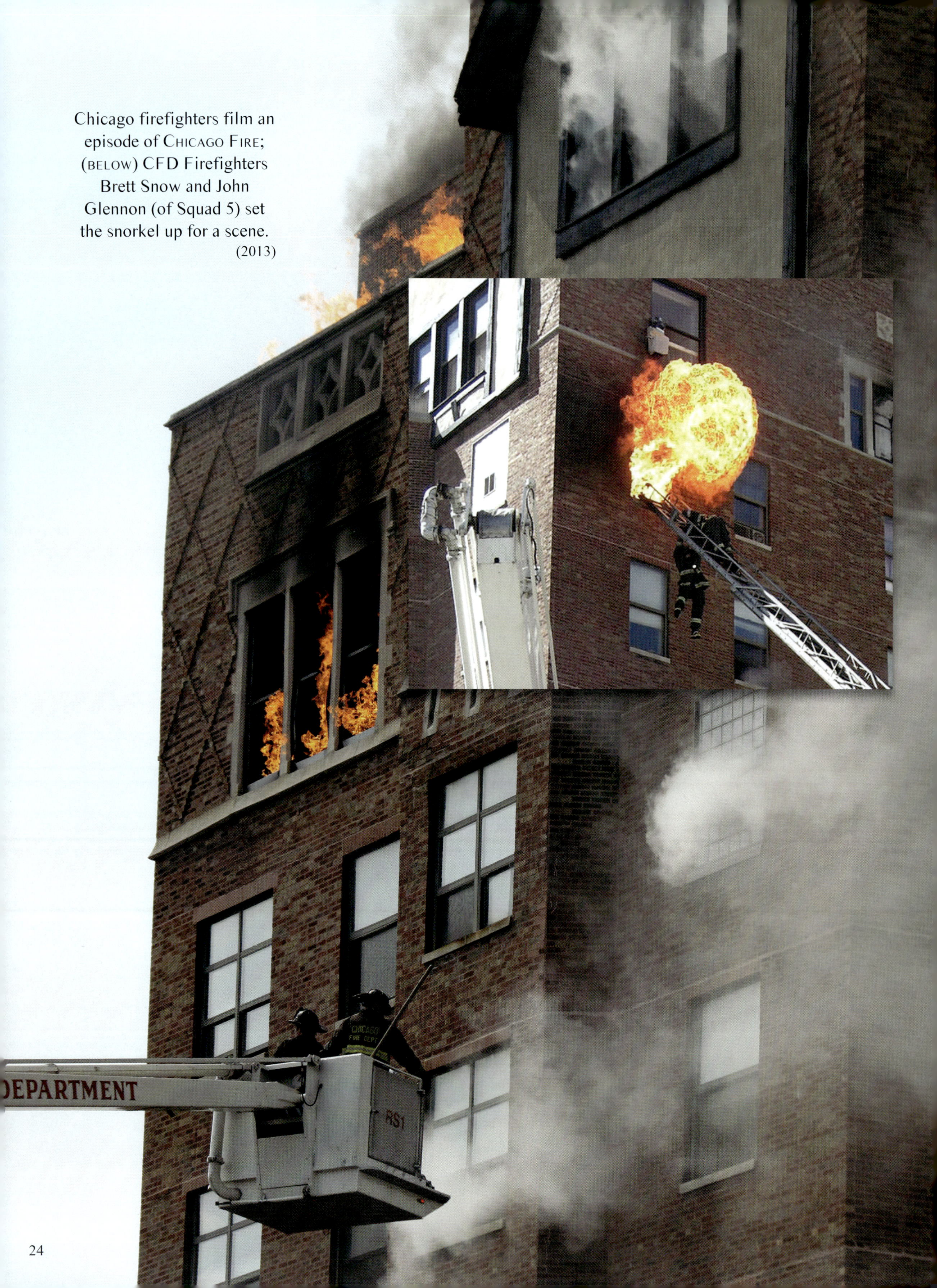

Chicago firefighters film an episode of CHICAGO FIRE; (BELOW) CFD Firefighters Brett Snow and John Glennon (of Squad 5) set the snorkel up for a scene. (2013)

LEFT: Captain Steve Rusin of the Nunda Fire Protection District watches the fire at Wildland Training with the Countryside Fire Department.

CENTER LEFT: Cicero Fire Department Truck 2 is covered in ice at an extra alarm house fire in Cicero during below zero temps.

Lyons Fire Chief Gordon Nord shows firefighters where he would like the tower set up at Broadview Fire Department's 2nd alarm fire at 1600 Roosevelt Road.

2009

Carol Stream Fire Chief Rick Kolomay and Deputy Chief Bob Hoff (2012)

Berwyn firefighters work to put out leftover fire on the roof of a house at 1310 East Avenue. (2012)

Chicago Fire Department members stand by on the movie set of the Batman movie—Firefighter Steve Scott, Chief Steve Chikerotis, Firefighter John Scheinpflug and Lieutenant Joe Kish. (2009)

2009

Chicago Fire Department Squad 1 member John "The Plug" Scheinpflug on the set of the Batman movie

CFD Fire Academy truck on the set of the Batman movie (2009)

Rockford Fire Department works at a vacant house fire in the south end of the city on June 23, 2012. The Blackhawk Fire Protection District arrived first and upgraded it to a 2nd alarm. (2012)

Chicago firefighters from Engine Company 91 prepare to extinguish an auto fire. (2012)

CFD officer opens the hood of the auto so they can find hidden fires. (2012)

Chicago Police Department pin-in accident on the north side at George & Sacramento

2013

Chicago Fire Department Squad 2 tries to free a man in a pin-in accident at Harrison & Kedzie involving a Chicago Police car.

North Maine Fire Protection District Chief Mike Fox (2012)

2013

2013

Chicago Fire Department 4-11 alarm warehouse fire at 21st & Western

Chicago Fire Department Engine 23 sprays water on hot spots at the ruins at 21st & Western.

2013

Chicago Fire Department Tower Ladder 5 works at th ruins the day after the fire at 21st & Western.

2013

Chicago Fire Department Tower Ladder 34 works at the 3-11 alarm apartment building fire on 111th Street. (2012)

Chicago Fire Department Squad 1 members Dave Leon, Joe Kubic, Kelly Burns, Tom Carbonneau and Jose Cotto

CFD Squad 5 Lieutenant Corey Hojek (2012)

Top: NIPSTA Fire Academy Instructors at live fire training at O'Hare Field

---

Center left: Chicago Fire Department Chiefs stop to take one of the last pictures with Retired Commissioner Bob Hoff; in the photo are Bob Hoff, Mike Fox, Bob McKee, The Bull, Tom Kennedy, Jeff Springer and Don Hroma.

RIGHT: Chicago Fire Department Squad 2 Firefighter Art Aranda (2008)

BELOW: Chicago Fire Department Squad 2 Captain Pat Maloney

2008

Chicago firefighters from Truck 35 and Squad 2 get off the roof at a 2-11 alarm fire at 3054 W. Armitage.

2008

Chicago Fire Department Tower Ladder 21 and Reserve Snorkel work to keep the fire from spreading to other buildings at the 5-11 alarm fire at Elston Avenue & Nelson.
(October 2012)

Chicago Firefighter Mike Curtain opens the windows on the second floor at a 2-11 alarm house fire.

2011

Firefighter Mike Curtain gets off the roof just as the fire blows out the windows that he just opened. (2011)

Chicago Fire Department 2-11 alarm and EMS Plan 1 Mayday house fire (2011)

A Chicago firefighter from Engine 109 is carried out after being injured at the 2-11 alarm fire. (2011)

A Chicago firefighter from Engine 39 shows off some ice that formed on his helmet at a 3-11 alarm fire on Morgan Street in December 2008.

LEFT: Zion Fire Department Firefighter Rocco Campanella works the aerial ladder to protect the exposures at the Maine Plastics plant fire at 1817 Kenosha Avenue.

BELOW: Winthrop Harbor Truck 1731 protects the exposures at a Zion Fire Department extra alarm fire at Maine Plastics at 1817 Kenosha Road.

Chicago Fire Department below grade rescue of a trapped worker who fell into a hole on the Dan Ryan Expressway at 47th Street (2006)

Chicago Fire Department Squad 2, 3rd shift firefighters—Firefighter Rowel, Firefighter Mendez, Firefighter Aranda, Captain Altman, Firefighter Velez, Firefighter Jurasz and Firefighter Whitley (2013)

Chicago Fire Department Senior Fire Alarm Operator and Fire Photographer Steve Redick (2007)

Chicago Fire Department Battalion Chief Lee Basile

RIGHT: Northfield Fire-Rescue Caption Tom Olk ventilates a roof at 4th & Maple in Wilmette (2009).

LEFT: Chicago Fire Department pumps foam into the basement of a bakery on Clark Street. (2007)

CENTER LEFT: CFD Squad 2 in the rear of the building at a 2-11 alarm fire at 7014 N. Clark

Chicago FD Foam Unit (2007)

2007

Retired Chicago Fire Commissioner Ray Orozco on the roof at the 2-11 alarm fire at 7014 N. Clark (2007)

LEFT: FDIC Pipes and Drums with Paul Teutul Sr. from Orange County Choppers

Cicero Fire Department Chief all iced up at an extra alarm house fire

Chicago Fire Department Truck 20 Lieutenant John Leonard (2012)

Chicago Fire Department divers search Lake Michigan for a missing man at North Avenue Beach.

2009

Chicago Fire Department divers and Chicago Police search for a missing man at North Avenue Beach. (2009)

Chicago Firefighter Joe Kubic
and Lieutenant Maibuecher

Chicago 5-11 Club members
Scott Stewart and Jim Mitidiero (2012)

CFD Squad 2 Captain Pat Maloney
and Firefighter Jim Reagan (2008)

Chicago Fire Boat 688
with the Chicago skyline
at North Avenue Beach
(2009)

2010

Chicago firefighters from Tower Ladder 39 vent the roof at 51st & Bishop

Chicago Fire Department 2-11 alarm fire at 51st & Bishop

2010

2012

Chicago Fire Department Battalion Chief Keith Witt

Chicago Fire Department Commissioner Ray Orozco and Deputy Fire Commissioner Bob Hoff at a 3-11 alarm fire on Devon Avenue, Thanksgiving 2007.

Chicago Fire Department still-and-box alarm fire at the silk flower warehouse at 4580 N. Elston. (2008)

Chicago Fire Department District Chief Rich Edgeworth (2008)

Chicago Fire Department Squad 2 Firefighter Alex Mendez (2008)

Chicago firefighters dig through the burned roofing at an extra alarm fire on Halsted north of Belmont. (2009)

CFD Engine 78 after the fire on Halsted north of Belmont

Chicago FD Lieutenant Pat Hannon at the 3-11 alarm fire at 3800 S. Morgan. Pat was a good friend of mine assigned to Truck Company 8 on S. Throop Street along with Engine Company 28. While on duty, Pat died in his sleep. Rest in peace, buddy; you are missed. (2008)

Chicago Fire Department Squad 1 Firefighters Joe Kubic and Dave Leon prepare to enter a pizza shop fire at Damen and Chicago Avenue.

Chicago Fire Commissioner Jose Santiago and District Chief Pat Donnelly

Chicago Fire Department firefighters try to free a man who is pinned in his car after being struck by an L-train on 4600 N. Kedzie Avenue.

Chicago Fire Department alley garage fire on Melrose (2008)

Chicago firefighters from Engine 91 try to protect the exposures from the alley garage fire on Melrose. (2008)

LEFT: Chicago Fire Department firefighters from Squad 5 recover a man who died in an accident at 103rd and the Bishop Ford. (2007)

BELOW: Chicago Fire Department District 6 Lieutenant Graves at a working fire in the 22nd Battalion at 115 S. Yale. (2013)

Chicago Fire Department Squad 2 after a 2-11 alarm fire at 6531 St. Louis in Lincolnwood on February 14, 2011, with Battalion 10 (2012)

Northbrook firefighters rescue kids trapped in a carnival ride that was stuck upside down.

2011

Glenview firefighters pose with a cardboard cutout of President Bush brought to the firehouse by a kid when the President flew into the Northeastern Illinois Public Safety Training Academy.

Chicago Fire Department Engine 18 at a 5-11 at 600 S. Wabash

Gurnee Fire Department working fire at the Gurnee Mills Mall (2008)

CFD Foam Unit works at the 3-11 alarm fire at Peerless Rug Company (2010)

Chicago Fire Department 3-11 alarm fire at the Peerless Rug Company at 3026 N. Lincoln Avenue (2010)

Glenview Fire Department 5-11 alarm fire at Red's Auto Body and Towing

2004

Northbrook Fire Department Retired Chief Jay Reardon at the 5-11 in Glenview

2004

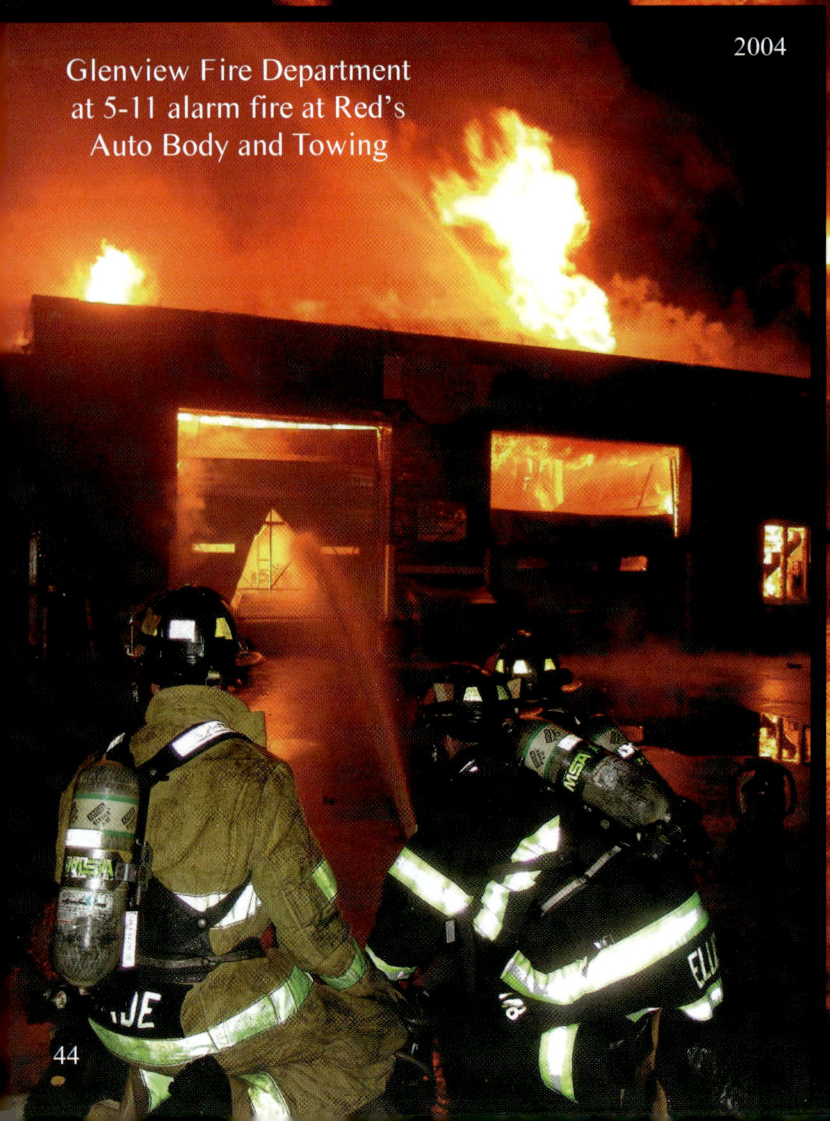

Glenview Fire Department at 5-11 alarm fire at Red's Auto Body and Towing

2004

Chicago Fire Department Squad 1
Firefighter Tony Budviatis

2012

Chicago Fire Department Chief Nielson,
Chief Niego, Chief Sampey and Chief Callahan

Chicago Tower Ladder firefighters open up the outside wall to gain access to a hidden fire. (2010)

Chicago firefighters torch a fence off its post so they can back a truck into the courtyard. (2012)

Cicero Fire Department

Chicago Fire Department
Chief Kevin Krasneck

Chicago firefighters work to free a man who jumped in front of a train at the Dan Ryan and 35th Street.

TOP LEFT: Chicago Fire Department Retired Fire Commissioner Ray Orozco

Left: Lyons Fire Department Chief Nord and Bedford Park Lieutenant Klotz (2011)

Photo at bottom of page: Waukegan Fire Department Crash Rig out of Waukegan Airport sprays water on a mock plane crash. (2012)

Chicago firefighters from Engine 78 work a line at a warehouse fire on the north side. (2012)

Chicago Fire Department
Battalion 7 Chief Rich Dory

2007

CFD District 5
Engineer
Pat Coogan
(2011)

CFD Engine 18
Firefighter
Claire Engel

Bensenville
Fire Department
Battalion Chief
Don Tessler and
son Matt (2011)

2011

RIGHT: CFD Training Officer at O'Hare Airport, Tom Wagner

Schaumburg Fire Department Rick Kolomay (2007)

Fire Engineering Editor and Chief Bobby Halton and Retired Fire Commissioner Bob Hoff (2012)

Deerfield Fire Department Battalion Chief Ken Koerber

Port Washington Fire Chief Mark Mitchell (2011)

Waukegan Fire Department extra alarm fire at the Salvation Army store (2012)

LEFT: Knollwood Fire Department Fire Chief Chuck Knesley and Chicago Fire Department Mike Mark at live fire training in Knollwood

Chicago Fire Department subway fire on Division east of Clark (2012)

Wheeling firefighter rescues a dog after a 2nd alarm fire in Riverwoods.

Chicago firefighters from Squad 7 drill on a plan prop. (2011)

Chicago Fire
Department
Chief Sampey
(2011)

Chicago Fire Department Fire Commissioner Bob Hoff and Deputy Fire Commissioner McNicholas carry the helmet of Christopher Wheatley, who passed away at a fire after falling from a fire escape.

Glencoe firefighter cuts a hood of a car during an extrication drill.

Chicago Fire Department Truck 20 rescues a dog off the roof from a fire.

Chicago Fire Department Special Operations Chief Mike Fox (2007)

Chicago Fire Department Truck 32 sets up at a 3-11 alarm fire at 3760 W. 38th Street. (2012)

Great Lakes Navy Engine 1912 pumps at the North Chicago Motel fire on Greenbay Road. (2012)

CFD Firefighter Kelly Burns

Northfield Fire-Rescue Firefighter Paramedic Sam Foster

Northbrook Fire Department Firefighter Paramedic Cliff Woodbury (2011)

Prospect Heights Fire Department Squad 9 pumps at an extra alarm fire at their City Hall. (2010)

Left: Glenview Fire Department Firefighter Paramedic Patrick Reed vents the roof at the Prospect Heights City Hall fire. (2010)

Bartlett Fire Department
train derailment (2011)

Chicago Fire Department Pipes and Drums at the memorial service at Rosehill Cemetery with the family of Lieutenant Pat Hannon (2012)

LEFT CENTER: Chicago Fire Department Local 2 annual memorial service at Rosehill Cemetery—Chief Nielson, Fire Commissioner Santiago, Chief Callahan, Chief Sampey, State Fire Marshall Larry Matkaitis, Chief McNicholas, Chief O'Connell and Chief Knapp (2012)

LEFT: Funeral for Des Plaines Fire Department Firefighter Tom Veverka in 2009

Chicago Fire Department
Level 2 hazardous materials
spill at 47th & Ashland.

Chicago Fire Department Engine Company 61,
located at 53rd & Wabash, is now out of service.

Waukegan Fire Department double
fatal house fire on Jackson

Alsip Fire Department
5-11 alarm lumber
yard fire (2007)

LEFT: Northfield firefighters at the Winnetka Drill Tower

Woodstock Fire Department live fire training (2009)

Lake Bluff Fire Department live fire training (2010)

Chicago Fire Department 2-11 alarm fire in the 2700 block of 79th Street (2013)

The Pink Heals fire truck JESSICA at the Schaumburg firefighters' Pink Tie Ball (2013)

RIGHT: Huntley Fire Department and Northern Illinois FOOLS member Shane Boehmke (2013)

Firefighters at an extra alarm fire at the meat packing plant in Cudahy, Wisconsin (2009)

Extra alarm fire in Cudahy, Wisconsin (2009)

Port Washington Tower Ladder sets up at the extra alarm meat packing plant fire in Cudahy, Wisconsin. (2009)

Chicago Fire Department Engine Company 93 at their firehouse the last day in ¾ boots before the switch to bunker pants

Chicago Fire Department Fire Commissioner Bob Hoff and Chicago Police Superintendent Garry McCarthy at the Battle of the Badges baseball game (2012)

Chicago Fire Commissioner Bob Hoff and son Andy at Cellular Field (2012)

New Lenox
Fire Department
live fire training

Lyons Fire Department Chief's helmet (2012)

Schaumburg
Firefighter
Jenna Werdell
(2012)

Northbrook Fire Department fatal car accident on Waukegan Road south of Techny Road. Here Northfield Fire-Rescue helps Northbrook with putting a fire out in one of the cars.

Deerfield Fire Department
live fire training

Knollwood Fire Department Retired Fire Chief Mike Issel and Countryside Fire Chief Jeff Steingart

Prospect Heights Fire Department Lieutenant Scott Olson at a working fire in Glenview (2013)

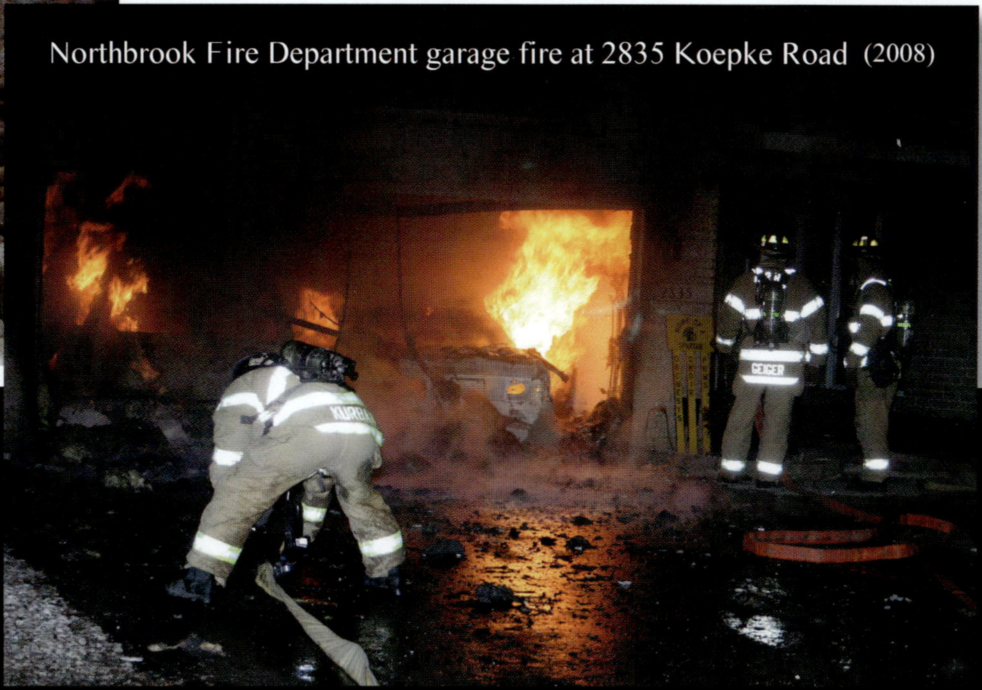

Northbrook Fire Department garage fire at 2835 Koepke Road (2008)

Northbrook Fire Department garage fire at 2835 Koepke Road (2008)

Northfield Fire-Rescue working fire (2007)

Northfield firefighters lead out on a working house fire which started in the garage. (2007)

Des Plaines Fire Department
Retired Battalion Chief Art Zern (2009)

Bristol Fire Department live fire training off Highway 50

Fox Lake Fire Department Brush Truck 2272 works at the wildland fire at Illinois Beach State Park.

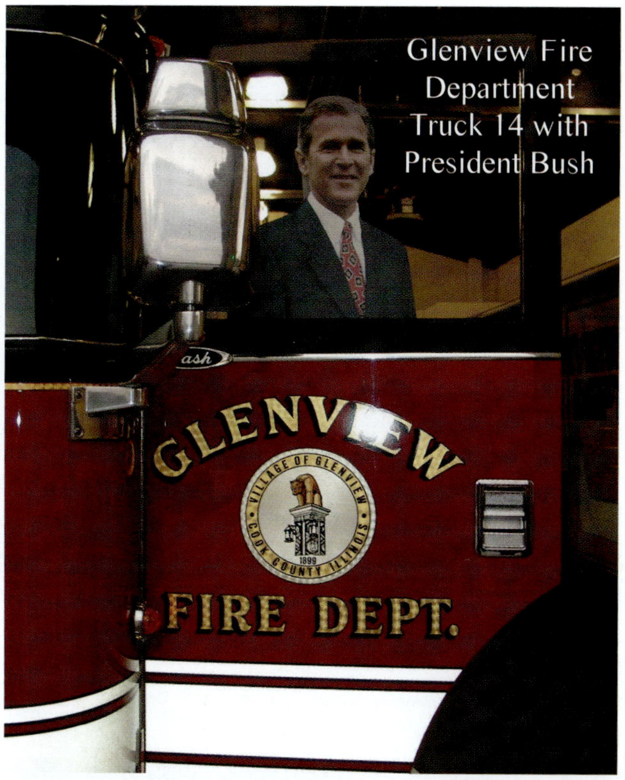

Glenview Fire Department Truck 14 with President Bush

Fox Lake Fire Department Engine 2222 at live fire training

Chicago Fire Department Fire Commissioner Jose Santiago (2008)

LEFT: Wheeling firefighters work to free a man from an accident on Dundee Road over the Tollway (2009)

Ashland Avenue 5-11 Fire, January 22-23, 2013

More than 170 firefighters battled a five-alarm blaze that engulfed an abandoned South Side warehouse and jumped to another structure. The building was at 3700 S. Ashland Avenue.

One firefighter suffered a minor back injury and was taken to Advocate Christ Medical Center in Oak Lawn.

The "extreme" fire was so extensive, emergency crews called two more special alarms, sending roughly one-third of the city's total firefighting power to help, officials said.

On Tuesday night, flames were "out of control" on all floors in the abandoned warehouse where the fire began. The roof and three walls of the 200-foot by 200-foot structure collapsed as crews fought the flames. The fire jumped to another building in

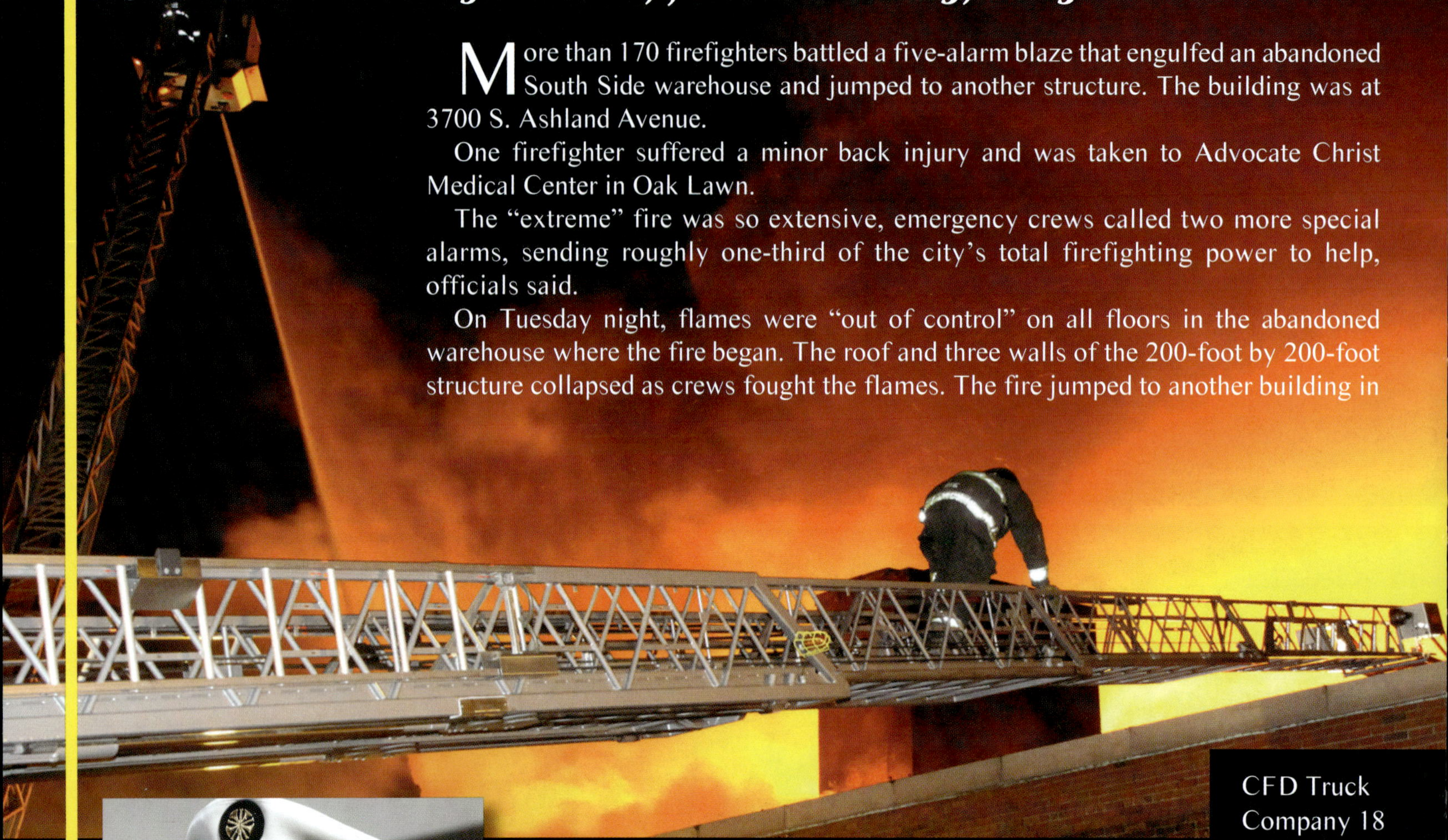

CFD Truck Company 18 working at the 5-11 alarm fire at 38th and Ashland.

Chicago Fire Department Fire Commissioner Jose Santiago

CFD Tower Ladder 39 keeps fire away from the flag company across the street from the Ashland Avenue warehouse fire.

A Chicago firefighter sits at the top of the ladder and waits for water.

CFD Engine 47 pumps at the 5-11 alarm fire.

CFD Tower Ladder 39 at the 5-11 alarm fire on Ashland Avenue

the 3800 block of S. Ashland Avenue, but firefighters halted the blaze from consuming that building.

In the past, homeless people had apparently started small fires in the building to keep warm. A firefighter who happened to be driving by the boarded-up warehouse first spotted the smoke coming from the vacant building and called it in to authorities.

The fire was particularly difficult to fight in 10-degree temperatures. Ice from freezing hose water glazed the brick exterior as the guts of the building burned. (Excerpted from the Chicago Sun-Times)

SNORKEL
CHICAGO FIRE DEPT.
FDG466
Seagrave

Chicago Fire Department Reserve Snorkel at the scene of the 5-11 on Ashland Avenue

Chicago Fire Department spare Tower Ladder E252, which used to be Tower Ladder 39, works the warehouse fire at 38th and Ashland Avenue.

CFD Deluge Wagon 671

CFD Commander Dan Swift, Engineer Ed Gil and Firefighter Mike DeJesus

Ashland Avenue 5-11 Fire, January 22-23, 2013

A truck covered in ice that was parked in the rear of the warehouse.

CFD Ed Gil and Commander Swift work Little Moe at the 5-11.

CFD Captain Kehoe watches his guys put water on the hot spots at the 5-11 alarm ruins.

CFD reserve snorkel at the ruins on Ashland

CFD Tower Ladder 39 works at the ruins on Ashland.

2007

Flight for Life leaves the scene of an auto accident with several injuries on the Tollway in Northbrook.

Flight for Life MedEvac takes off over the Mess Canteen at the Lake County Fire Chiefs Expo at Gurnee Mills Mall. (2008)

Glenview firefighters remove an injured worker from a hole that he fell in at the Glenview water treatment plant.

Chicago Fire Department Chinatown Engine 8

RIGHT: Chicago Fire Department Snorkel Squad 1A

Lincolnwood Fire Department fatal car accident on Cicero north of Pratt

Chicago Fire Department Retired Battalion Chief Dory (2005)

North Shore, Wisconsin Fire Department ladder lights up the scene at a house explosion. (2012)

Northbrook Fire Department Battalion Chief Dan Quinn (2012)

Indianapolis Police show Bob Welch who is boss at FDIC.

2012

Chief Billy Goldfeder and Sarah Dagdick and her father, Captain Ed Dagdick of the Wauconda Fire District

Chief Billy Goldfeder and Chief Dave Traiforos at FDIC in Indianapolis

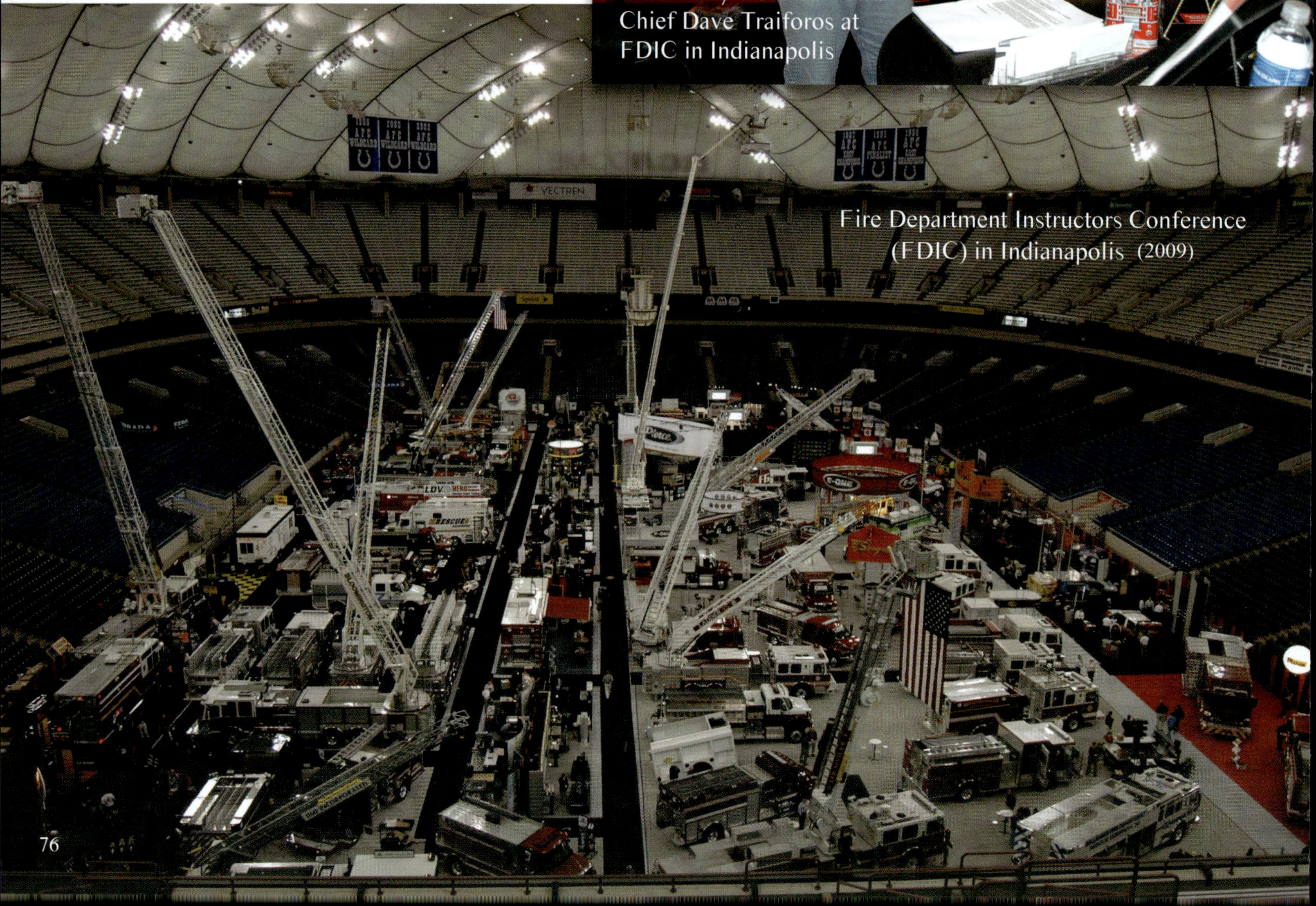

Fire Department Instructors Conference (FDIC) in Indianapolis (2009)

Battalion Chief Rich Edgeworth gets a little line time at a Chicago Fire Department still and box alarm fire at 70th and Parnell. (2005)

Northbrook Fire Department Chief Mark Nolan (2012)

Evanston Engine 24 tries to keep the fire from spreading at an extra alarm junk yard fire. (2007)

Evanston Fire Department 2nd alarm fire at North Shore Towing Auto Salvage Yard (2007)

Below: Chicago Fire Department Retired Commissioner Bob Hoff and new Commissioner Jose Santiago (2012)

Chicago Fire Department Firefighter Larry McCormack, the FDIC 2012 FDNY Ray Downey Courage and Valor Award winner. Pictured in this photo with Larry are Ray's sons, Joe and Chuck Downey. (2012)

Chicago Fire Department Retired Chief Ray Hoff

FDIC Pipes and Drums and the girls from the Indianapolis Hooters (2009)

FDNY Bronx Lieutenant Mike Ciampo and his students remember Dana Hannon at FDIC Hot Classes (2012)

Waukegan Fire Department runs the Firefighter Combat Challenge in Oak Forest

BELOW: NIPSTA Instructor Dave Traiforos shows off his new shield

Chicago firefighters from Squad 2 breach a basement wall to get to a basement fire in Lincolnwood.

Elk Grove Village extra alarm house fire

Chicago Fire Department Lieutenant LaHendro and his son Chris

ABOVE LEFT CENTER: Flight for Life lands on Greenbay Road in Waukegan. Gurnee engine sets up the landing zone south of the accident.

RIGHT: Elk Grove Village Engine 7

Glenview Truck 14 at the Glenview Park District maintenance building fire

Monroe Fire School Instructor Dave Traiforos teaches live fire training.

2012

Glencoe Public Safety Lieutenant Betsy Seno at LP gas live fire training (2012)

Live fire training at FDIC Hot Classes (2008)

Heavy black smoke comes from a house before companies put water on it.

2011

Gurnee Fire Department firefighters fight a mobile home fire. (2011)

Highland Park Firefighter Ryan Hughes at an extra alarm fire on Sheridan Road (2011)

Chicago Fire Department Retired Chief of 30 years Andy O'Donnell (2012)

Evanston Captain Guy Boliaux

Gurnee Fire Department mobile home park fire on the east side of town (2011)

Glenview firefighters work to gain access to a garage so they can put out the car that is on fire inside.

Franklin Park Lieutenant Doug Halverson (2013)

Chicago Fire Department Engine 62 and Truck 27 at a still and box alarm fire on the south side. (2012)

Winnetka firefighters pull a ceiling at a house fire in Wilmette. (2010)

Chicago Fire Department Squad 1—Firefighter Owsianiak, Firefighter Burns, Firefighter Kubic, Firefighter Mielnicki, Captain John Collins and Firefighter Leon (2006)

Des Plaines Firefighter Milton after a factory fire in Des Plaines

2011

NIPSTA Instructors Sam Foster, Pat Schey and Ron Moritz (2011)

Chicago Fire Department Captain Kehoe and Deputy District Chief Steve Chikerotis (2012)

Long Grove firefighters advance hose line to the west side of a house to keep the fire from spreading

Lake Forest Deputy Chief Chris Garrison and Countryside Deputy Chief Kris Kazian

Hebron Fire Department live fire training (2012)

Chicago Squad 1 firefighters after a still in an apartment building—Firefighter Dave Leon, Firefighter John Scheurich, Firefighter Jim Stepien and Firefighter Kelly Burns

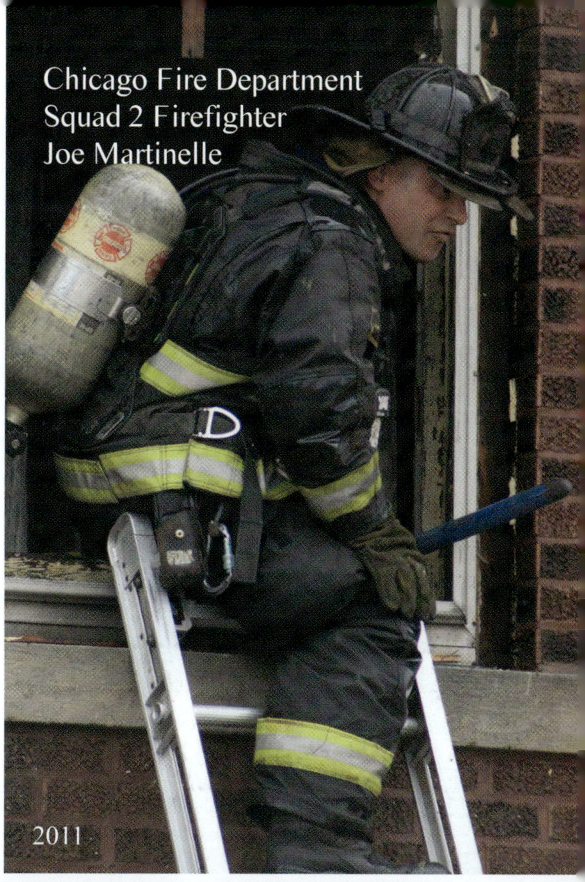

Chicago Fire Department Squad 2 Firefighter Joe Martinelle

2011

Lincolnshire-Riverwoods fatal plane crash (2012)

New Lenox Fire Department live fire training

Air One police rescue team (2012)

Chicago Fire Department Battalion Chief Rich Dory and Battalion Chief Jerry McKee

Chicago Fire Department Lieutenant Panatera, Firefighter Clemens, Battalion Chief Burke and Lieutenant Leonard (2012)

Knollwood Fire Department Captain Bill Knesley and his nephew Joe Knesley (2009)

Chicago Fire Department 2-11 alarm fire at 21st & Western. My nephew Thomas and I chased this fire and met up with long-time friend Lieutenant Pat Hannon of Truck Company 8.

2010

Long Grove Fire Department box alarm house fire

Northbrook Fire Department house fire in the 1500 block of Shermer Road (2012)

Northbrook Deputy Chief Bob Brogan talks with Lieutenant Jeff Harris and Northfield Assistant Chief Tom Burke at the scene of a working house fire on Shermer Road.

2012

McHenry County College Firefighter 2 Class of 2013

2012

Lake Forest firefighters work a hand line from the roof of a fire in Lincolnshire.

2007

Lake Bluff Fire Department Engine 517 pumps at their drill tower.

Office of Emergency Management's Shamarr Thomas and Chicago Fire Department Squad 5 Firefighter Mike Sullivan

Lincolnshire-Riverwoods Chief Fred Kruger (2007)

Minooka Fire Department tanker pumps water to Joliet's tower ladder at a 5-11 alarm casino fire.

Newport Fire Department
7 alarm house fire off
Wadsworth Road
(2008)

Newport Fire Department
Lieutenant Rob Schreiner
at the 7 alarm house fire
in Wadsworth (2008)

Loyola Lifestar prepares a football player with a neck injury to be flown to Luthern General Hospital.

Lake Forest ambulance after
a traumatic arrest call with
Flight for Life

MABAS Divisions 1, 10
and 27 new Tempest
Mobile Ventilation Units
(2012)

2012

Lincolnshire-
Riverwoods 3-11
alarm house fire
(2012)

Loyola Lifestar

2012

Firefighters pump foam into the basement of a house fire in Riverwoods

Top left: Northbrook Fire Department Lieutenant Steve Johnson's last day at the 3-11 alarm house fire in Riverwoods

Right: The monkey bottle change

2013

Left: Chicago Fire Department Squad 1 member Justin Reynolds (2012)

Center left: Chicago Fire Department Squad 1 Firefighter Jim Sewnig

Knollwood Firefighter Kerry McWilliams rolls hose at a north Chicago box alarm. (2013)

Lake Forest Fire Department accident with a man pinned in a car—special call for Flight for Life

Northbrook Fire Department
Chaplain Aron Melman (2012)

Lombard firefighters ice rescue class (2012)

Above: Lake and McHenry County Side-Scan Sonar Team

McHenry Fire Department
dive call with the side-
scan sonar team (2012)

Lincolnshire-Riverwoods new Truck 53 designed by the firefighters (2012)

BELOW: Antioch tower ladder works at a fire in a plastics plant in Zion

2012

Milwaukee Fire Department Engine 18

Chicago Fire Department 2-11 alarm fire in Chinatown

Chicago Fire Department Squad 1 third shift—Firefighter Truesdale, Firefighter Budviatis, Lieutenant Duffy, Firefighter Heenan, Firefighter Dyer and Firefighter Curtin

2011

Lake Bluff Engine 519 (2012)

Joliet Fire Department 5th alarm fire at the Empress Casino on March 20, 2009

Zion Fire Department
5-11 alarm plastics plant fire
(2012)

2010

Firefighters apply foam to a rolled over tanker truck while a worker drills holes to pump out the truck. (2010)

Morton Grove firefighters protect the worker who is drilling holes to pump out the truck.

2010

2010

Glenview Battalion Chief Spencer Kimura and Highland Park Fire Department's Charlie Orth at the scene of a rolled over tanker truck

Lin-Mar Towing and Recovery rolls the truck upright after it was pumped out. (2010)

TOP LEFT: Chicago Fire Department Chief Sampey talks over the plan of attack with Haz-Mat Ops Officer Tom Kanelos on a leaking tanker truck that rolled over.

Chicago Fire Department
Fire Boat Engine 58 (2011)

Chicago firefighters cool off the bridge at 95 and the canal; the bridge wouldn't close due to the heat. (2011)

Palos Heights firefighter checks for fire extension (2012)

Lake Forest Fire Chief Jeff Howell and his brother, Lombard Fire Department Deputy Chief Jerry Howell (2012)

Northbrook Fire Department Lieutenant Steve Johnson rescues a cat from a burning house.

2011

Huntley Fire Department fatal apartment fire

Northbrook Fire Department double fatal train derailment with bridge collapse

Northbrook Fire Department train derailment (2012)

Northbrook Fire Department
Lieutenant Jim Bullock (2012)

2011
Mount Prospect Fire Department
fatal accident on Mount Prospect
Road and Northwest Highway

Mount Prospect Fire Department
train-vs.-semi fatal accident (2011)

North Palos
house fire
(2013)

Northbrook Fire
Lieutenant Jim Bullock

2011

Evergreen Park
Assistant Chief
Corey Hojek
(2012)

Fire Department Photo Unit
Scott Stewart (2012)

Wheaton Fire Department
Shift Commander Lou Lake (2010)

Wheeling Fire Department
fatal confined space incident

Chicago Fire Department
Air Sea Rescue (2012)

North Shore Fire Department
3rd alarm house fire (2012)

North Shore Fire Department
3rd alarm house fire (2012)

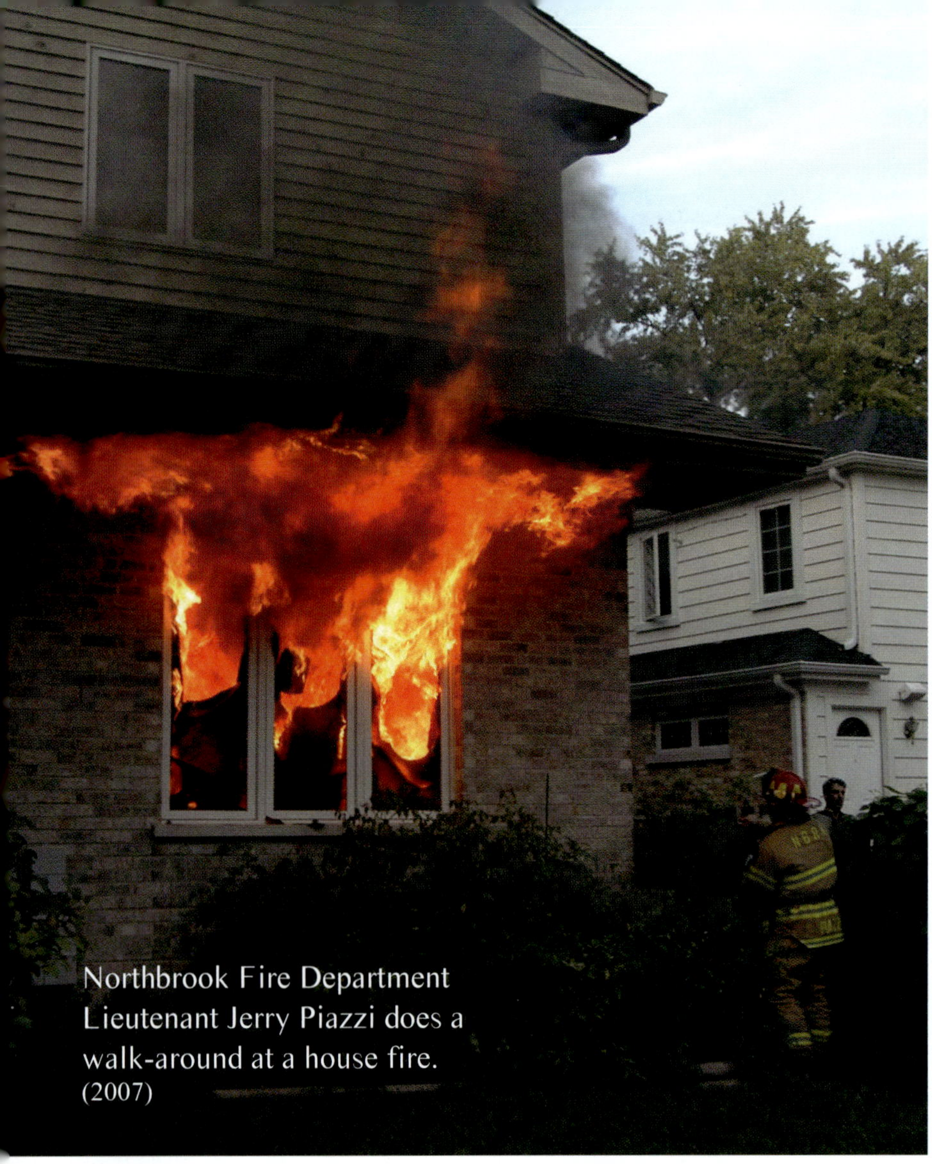

Northbrook Fire Department Lieutenant Jerry Piazzi does a walk-around at a house fire. (2007)

Northbrook firefighters advance a 2½-inch line into a house with heavy fire in the rear. (2007)

Chicago Fire Department Battalion 22 and Engine 62 and Truck 27 (2013)

Lake County MABAS Division 4 Specialized Response Team (2008)

Lake Villa Fire Department Firefighter Rollene (2008)

Mendota Tower Ladder 617 (2012)

Mendota Fire Department Tower Ladder works at the Peru Clock Factory fire.

2012

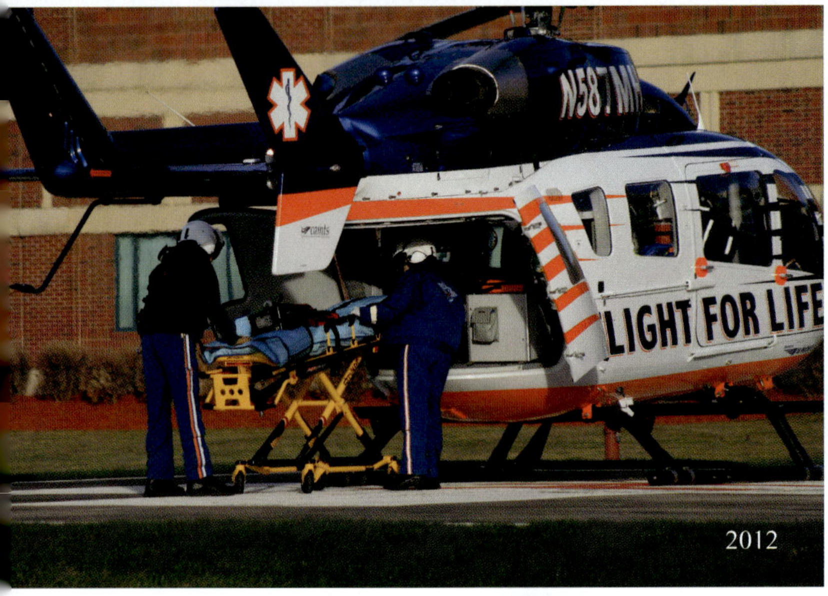

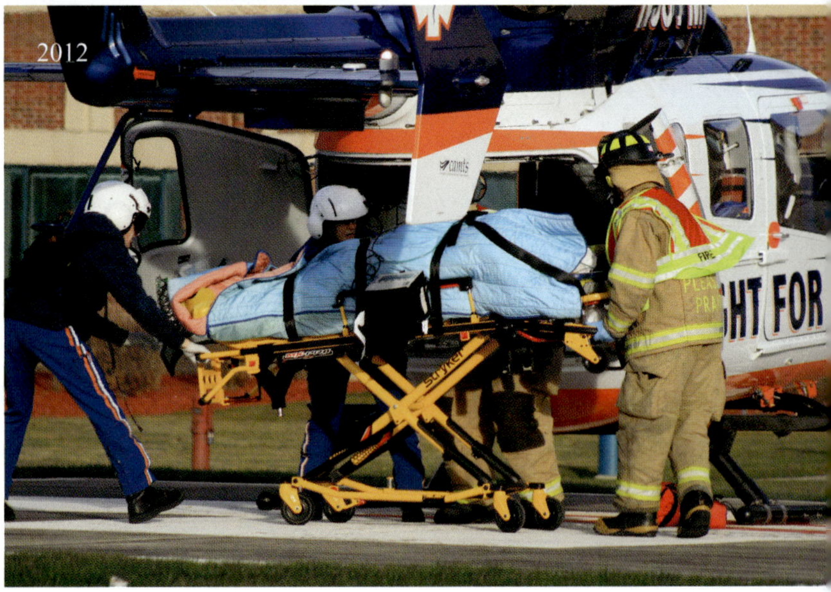

Top: Three freight trains derail in a cornfield in a rural area of Indiana.

Center, left & right: Flight for Life

Right: Flight for Life takes off with a person who was trapped in a car in Pleasant Prairie.

LEFT: Northbrook Fire Department multiple semi accident on the Edens Spur (2011)

CENTER LEFT: Ernie's Wrecker Service in Vernon Hills works at the double semi accident on the Edens Spur in Northbrook (2011)

Wauconda Fire Explorer Grace Abernathy (2011)

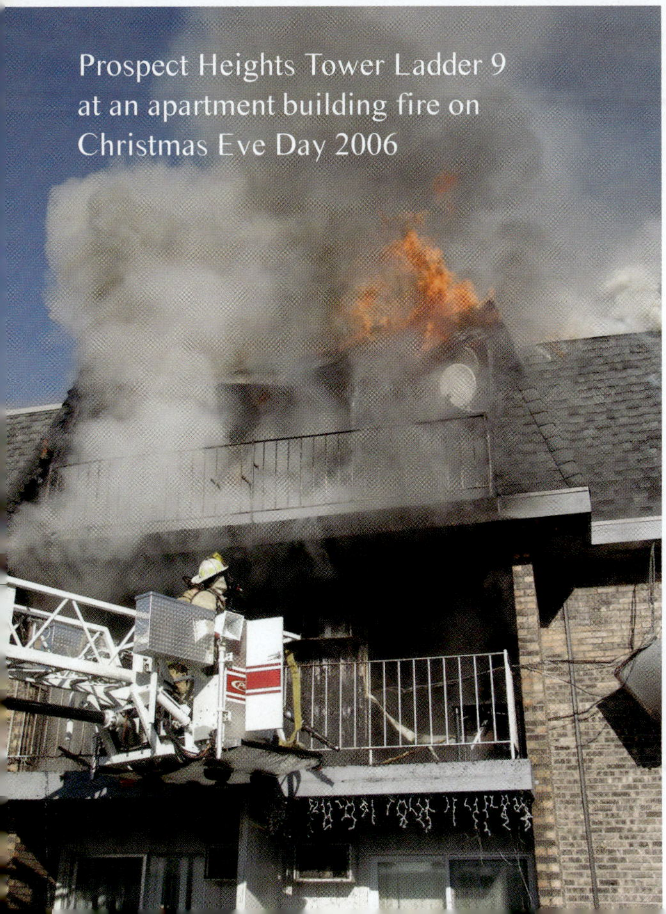

Prospect Heights Tower Ladder 9 at an apartment building fire on Christmas Eve Day 2006

Prospect Heights 4-11 fire (2006)

Spring Grove Fire Department live fire training

2007

Woodstock Fire Department live fire training

2012

Chicago Fire Department Squad 7 firefighters vent a roof in Bensenville at a live fire training. (2011)

Bristol auto accident on Highway 50 and Route 45 (2011)

Bristol Fire Department pin-in accident at Route 45 and Highway 50 (2011)

Flight for Life lands in Bristol, Wisconsin

2011

Bristol Fire Department firefighters load a woman who was pinned in her car in an auto accident

2011

Wheaton Fire Department roof fire at Briar Glen School

2012

Skokie Fire Department
working house fire (2007)

Skokie Lieutenant Mark Reding helps clean out the doorway so firefighters can get to the fire. (2007)

Skokie fire Department
working fire (2007)

Carol Stream
Fire Chief
Rick Kolomay
(2012)

Right: Downers Grove Fire Department pin-in accident

Center right: Northbrook firefighters fight a tar kettle fire next to a house

Below: Winfield firefighters vent a roof at a working fire

Heavy smoke from a house fire just before firefighters arrive (2011)

Fire vents from the second floor window (2011)

Wilmette firefighters
check for hot spots
(2012)

Evanston Fire rescue divers search
Lake Michigan for a missing boy (2012)

Northbrook Fire Department pin-in accident on the Edens Spur

Chicago Fire Department Bobby Hoff with Firefighters Hojek, Cogsgrove and McArdle and Lieutenant Will Trezek

Center right: Wilmette Fire Department auto accident with six people trapped in a car after a tree falls on it

Right: CFD Squad 2 Firefighter Joe Martinelle

Evergreen Park Fire Department Snorkel Squad 3

Newport Fire Chief Mark Kirschhoffer backs up Firefighter Taylor at a fire in Zion.

Heavy fire from a second story window in Wilmette

Winnetka firefighters fight a Wilmette house fire.

2013

Northbrook Fire Department firefighters from gold shift at Station 10—Retired Deputy Chief Ron Schinleber, Retired Firefighter Hal Sanger, Firefighter Thom Kanelos, Firefighter Chris Gray, Deputy Chief Tim Smeltzer and Retired Firefighter Ray Starzec

Northbrook Fire Department retired member Hal Sanger

2010

CFD Squad 2 Firefighter Bryan Velez (2012)

RIGHT: CFD member Justin Reynolds talks over the live fire training at Monroe Fire School with the Live Fire Safety Officer.

2013

2013

2013

CHICAGO
FIRE DEPT.

NORTHBROOK F.D.

Knesley family, Mother's Day 2002

A6
GLENVIEW
A8
GLENVIEW

DEERFIELD
62
BANNOCKBURN

Index